YOUR KNOWLEDGE HAS VALUE

- We will publish your bachelor's and master's thesis, essays and papers

- Your own eBook and book - sold worldwide in all relevant shops

- Earn money with each sale

Upload your text at www.GRIN.com and publish for free

Bibliographic information published by the German National Library:

The German National Library lists this publication in the National Bibliography; detailed bibliographic data are available on the Internet at http://dnb.dnb.de .

This book is copyright material and must not be copied, reproduced, transferred, distributed, leased, licensed or publicly performed or used in any way except as specifically permitted in writing by the publishers, as allowed under the terms and conditions under which it was purchased or as strictly permitted by applicable copyright law. Any unauthorized distribution or use of this text may be a direct infringement of the author s and publisher s rights and those responsible may be liable in law accordingly.

Imprint:

Copyright © 2015 GRIN Verlag
Print and binding: Books on Demand GmbH, Norderstedt Germany
ISBN: 9783668536449

This book at GRIN:

https://www.grin.com/document/372957

Teresa Ruß

Immigration in London. The cultural Hotspot of England

GRIN Verlag

GRIN - Your knowledge has value

Since its foundation in 1998, GRIN has specialized in publishing academic texts by students, college teachers and other academics as e-book and printed book. The website www.grin.com is an ideal platform for presenting term papers, final papers, scientific essays, dissertations and specialist books.

Visit us on the internet:

http://www.grin.com/

http://www.facebook.com/grincom

http://www.twitter.com/grin_com

Introduction

Imagine walking around in London – but not seeing the ethnic groups from all over the world, with their international restaurants. Imagine London without Brick Lane Market or the various food stalls at Camden. This can't be the London that we all know, right? So where does this cultural variety come from? The answer lies in Immigration, which is an very important factor for life in London and the United Kingdom. Almost every British city serves as a new home for Immigrants but none quite reaches the extent of London. A statistic from the Migration Observatory in the United Kingdom from 2012 shows that Inner and Outer London are definitely the most important destinations for Immigrants in England. In the centre of London around 40 % of the population is foreign born and 24 % are foreign citizens. These numbers seem to be huge, especially in comparison with another big city like Manchester, where only 6 % of the population are foreigners.

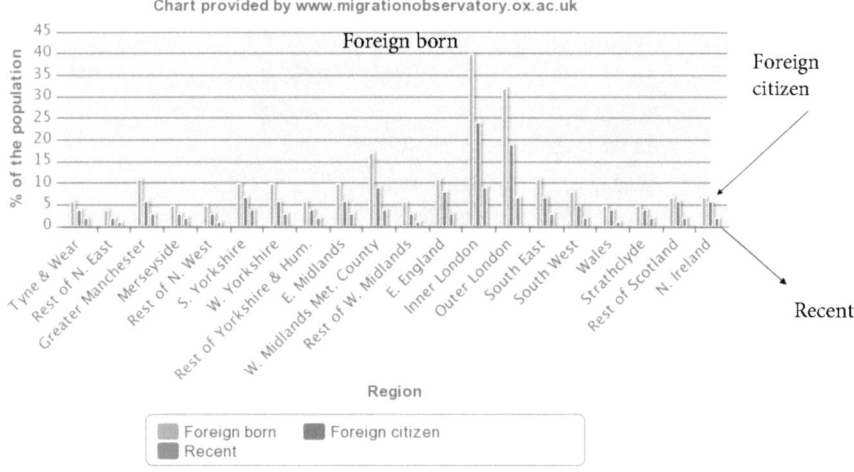

(http://bit.ly/1jcSkTu , 26-10-2014)

So these were just some little facts and figures to start with. The next pages will give you a good overview about the History of several Ethnicities in London and how they shaped the Culture by bringing their Traditions and Culture into the Country.

1. The Immigration History of different ethnic Groups in London

1.1 Asians

1.1.1 Immigrants from the Indian Subcontinent

Indians represent the biggest ethnic minority in London, as underlined by recent statistics:

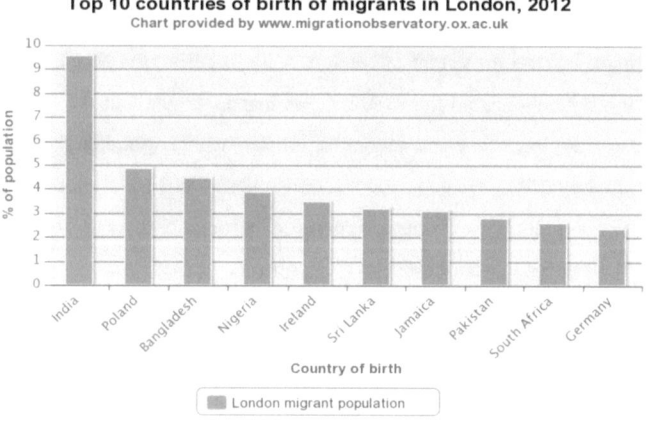

(http://bit.ly/1jcSkTu , 26-10-2014)

Roughly 10 % of the population in London consists of Indians. In this case, Indians means not only people from India itself, but also from Bangladesh, Pakistan and Sri Lanka.

The major wave of Immigrants from the Indian Subcontinent started coming into the Country during the 1950s and 1960s. One reason of this rapid rising was the "British Nationality Act" in 1948, which allowed citizens from the British Commonwealth free entry into Great Britain whilst the United States introduced a stern new law concerning the Immigration policy. This made the United Kingdom even more attractive for Foreigners. Another reason was that the Immigrants soon noticed that Britain was in desperate need for new workers after the destruction of the second World War [1]

---

[1] http://bit.ly/1nK9len, 27-10-2014

Probably the most important conjunction between India and Great Britain is and always will be the British Empire with its colonies.

> *"In 1900 it was not absurd to regard London as the centre of the world, and children learned certain phrases which expressed in simple terms the 'truths' which the British regarded as paramount: The sun never sets on the British Empire; India is the brightest jewel in the imperial Crown; Britannia rules the waves."*[2]

The British started to gain power in the country in the 17th century.[3] But India wanted to have Independence again, so in 1947 the British Rulers in India had to give in and India became a republic while still remaining part of the Commonwealth.[4] The young people from India seized their opportunity and came to Britain to attend universities and colleges. Since the people from the Commonwealth had free entry to Britain, the country was faced with a wave of Immigrants who thought that Life in the welfare state Britain would be better than to remain in their mother country where living conditions still remained fairly low. In 1961, immigration from the Commonwealth was rising so fast, that the British Government was forced to introduce laws to restrict it.[5]

But nevertheless, once they were there they couldn't be forced to leave the country again and go back. This is why London has got a large Indian community especially focused in London's East and West End.[6]

(a bilingual street sign in London's East End)

---

2 Bromhead, *Life in Modern Britain*, p.216
3 cf. http://bbc.in/KwQIbG (02-11-2014)
4 cf. Bromhead, pp. 217-220
5 cf. ibid. p.221
6 cf. http://bit.ly/1s4Lq5O (02-11-2014)

1.1.2 Chinese

Thinking about the Chinese population in London eventually leads to the image of Chinatown in Soho. But this Chinatown is not the first ever to exist in London. The first Chinatown was in the Docks, where the Chinese had their businesses to supply the sailors. But this area was destroyed during the Second World War. The result is the present Chinatown in Soho, which was established in the 1970s. Most tourists would think that this is where the Chinese community lives, but there's more retail than residents there. Meaning there are a lot of restaurants and shops in Chinatown, but not the living space of the owners. [7]

(http://bit.ly/1qlDiNM, 02-11-2014- New Year's preparation in Chinatown, London)
The Chinese population in London at present time is not focused on one particular part of the city but distributed evenly on the boroughs of the city. A lot of Immigrants from China who are now owners of a business in Chinatown came in the 1960s, because Chinese restaurants experienced a boom and they had great success. Vietnamese fugitives escaped the war in the 70s and settled in Lewisham, Lambeth and Hackney. Ten years later a lot of well educated people from China migrated to the London suburbs of Croydon and Colindale.[8] Recent figures show that the most new immigrants come from China, the number rose drastically from 2008 to 2012, which can be lead back to the high number of young Chinese people attending universities in Great Britain.[9] The good connection between Britain and China can be seen in the celebrations for the Chinese New Year on Trafalgar Square.

7 cf. http://bit.ly/1A0iDlt (02-11-2014)
8 cf. http://bbc.in/1tQvf0h (02-11-2014)
9 cf. http://bit.ly/1vV5wTI (02-11-2014)

1.2 Africans

The History of Africans migrating began in the 1500s when the slave trade began - they were not only needed to work on plantations, but also in the wealthy English households. At the beginning of the eighteenth century, with the Industrial Revolution just about to start, the rich British merchants from the big cities brought back not only goods, but also even more African slaves. An estimated 14.000 black women, men and children lived in England in 1770. 1807 was the year the parliament banned the trade of slaves, but not slavery itself. So, the slaving trade proceeded and when captains thought they would be caught they threw the slaves overboard. Almost 30 years later, in 1833, the parliament finally banned slavery all across the British Empire. When Immigration from Europe was increasing, the Arrival of African People almost stopped. The only black immigrants at that time were Seamen that settled down with small communities near the British harbours. Restrictions on Immigration by the government made in the 1970s made it much harder for black people to migrate than for white. In the 1980s only specific groups migrated because of the stricter laws. South Africans for example made use of the "family-tie" entry rule, which means that they had relatives already living in Britain and followed them. In the following years, a lot of people suffered from political or religious persecution and searched for asylum. In the two years between 1998 and 2000, 45.000 asylum seekers arrived from Africa alone. [10]

Many of the African immigrants nowadays do not come from Africa directly. They first migrate to another European country, say Germany or Sweden and then decide to go to Britain. The reasons are that they think the United Kingdom is more open for Immigrants than other European Countries - perhaps because of its centuries-long Immigration history? They also admire that so many non-white people succeed in Britain, especially in London and hope to do the same.[11] Unsurprisingly that Peckham has got the largest Nigerian community in Britain, although some other popular African areas in London include Hackney, Lambeth or Southwark.[12]

---

10 cf. http://bbc.in/1yONYK6 (02-11-2014)
11 cf. http://bit.ly/1t36LMa (02-11-2014)
12 cf. http://bit.ly/1xPwDzl (02-11-2014)

1.3 Immigrants from Europe

1.3.1 The Irish

Being the Neighbour of England and Northern Ireland one part of the United Kingdom, not often do we think about the Irish when it comes to the topic of Immigration to London and Britain. But in a 2001 study it says that 24% of the British population have an Irish parent or grandparent, which makes a total of 14 million. In comparison to a census from 1991 according to which 10% have Irish ancestors, the numbers have been rising vigorously.[13] These are the people who had Immigrants as parents or grandparents. But what about the people born in Ireland and immigrating to Britain? In 2001, 494.850 people who were born in the Republic of Ireland lived in Britain. 157.556 of these people alone lived in London.[14]

But why is that? Robert Winder states it like that:

> "By far the biggest injection of fresh blood in the nineteenth century came from Ireland, and the fury with which Britain reacted is famous. Few Immigrants have been less welcome.(...)They were penniless, unhealthy, unshod and unclean, lacking even the wherewithal to wash."[15]

The Irish Migrants always did the lowest-paying jobs. Whether it was digging Canals, picking Fruits or hacking Stones. It has always been the Irish who did the jobs the 'normal' English population would never do. There even were some riots from the English labourers, who got fired for the much cheaper Irish workers by their employers. In London, the Irish rioted then again against the nasty working conditions in the docks. These Irish workers, who did almost everything for a bit of money hoped to earn something to bring back to their families .[16]

In the 1840s the huge problem of Starvation occurred in Ireland. Some managed to go to America and even were valued for their tough way of working. But those who took the shorter way to England would soon notice that they weren't better off with that choice. [17]

---

13 cf. http://bit.ly/1wUCuW0 (02-112014)
14 cf. http://bbc.in/13vrpju (02-11-2014)
15 Winder, *Bloody Foreigners*, p.194
16 cf. ibid. p.195
17 cf. ibid. p.196

Before doing the work none of the English would do, in 1861 the Irish were not even immigrants, but they were British. (Until 1922, when the Republic of Ireland went independent.)[18] Still, the British thought of their fellow countrymen in all the bad ways one could, they thought of every cliché one could have about the Irish:

> "Far from being greeted as fellow-nationals, they were seen as barbaric, more or less. They were poor, uneducated, didn't mind a drink and occasionally expressed themselves with their fists. As usual, they were seen as rebellious by nature and simply not to be trusted."[19]

So they treated them like Immigrants nonetheless because they didn't really speak English and they were Catholics, for example.[20] They were even prosecuted for their religion. And if that wasn't enough they had to deal with the poorest living standards:

> "Like many other groups of social unfortunates, the squalor in which they were trapped was held against them, as if they wore rags and lived in overcrowded slums by choice, as if they ate discarded turnip heads or potato peel because they thought them delicacies."[21]

This one sentence shows the disastrous living conditions Irish people had to deal with. No money to buy proper clothing or food and living together on smallest space in slums. (for example the settlements in St. Giles in the fields.) The consequence of these circumstances was an ever-growing fatality rate amongst the Irish.[22] They had to suffer so much and still they did a lot to build the country we know nowadays. Ironically, even though Irish Migration has strikingly decreased the last 20 years, a lot of Brits have got Irish Ancestors. Also, the situation for Ireland has changed:

> "Today, Ireland, one of the first modern nations to witness mass emigration to Britain, North America and the Australias, is today itself becoming a nation attracting immigrants." [23]

---

18 cf. Winder, p.197
19 ibid. p. 197
20 cf. ibid. p.197
21 ibid. p.197-198
22 cf. http://bit.ly/1wV04lo (02-11-2014)
23 http://bbc.in/13vrpju (02-11-2014)

1.3.2 Jews from East Europe

Jewish Immigrants from Eastern Europe, the Ashkenazi Jews came to London at the end of the 18th Century. Some were lucky and established a good business with banking, but others were not so fortunate and had to sell ribbons, watches, jewellery or clothes to survive. The Jewish population in Britain grew in the late 18th and the beginning of the 19th century, as well as their reputation. The families became wealthier and better educated.[24]

One success story that began with a Jew fleeing from Russia to avoid anti-Semitic campaigns was the story of Marks & Spencer, one of the most famous department store chains all over the United Kingdom. Michael Marks was leaving his home town in Polish Russia and arrived in England without speaking English or any money. When he opened a stall in the market in Leeds, he soon noticed that his idea was so good that he had to expand. He searched for a partner and found him in Thomas Spencer – Marks & Spencer was born. He wasn't the only one who came to succeed in Britain. The refugees from Eastern Europe often dreamed to go to America, but many simply did not have the money or the power to face such a journey. So they chose England, which was not only nearer, but also had an established Jewish community. But not alone did persecution drive the Jews out of the Land, in 1866 Russian Poland suffered from a Cholera epidemic and Lithuania from famine. Both catastrophes lead to a vast number of newly migrating Jews from Eastern Europe. The Jews occupied the despised areas of the docks, and there were a lot of them. Between 1881 and 1914 about 150.000 Jews came to England to start a new life.[25] They crowded together in the hope that they would be safe protecting each other and being ignored from the rest of the population. They isolated themselves from the rest of the population by different looks, music, food and shops. With the little English they spoke, they often worked in factories or sweatshops.[26]

But still, the Jewish Community brought a lot of culture and religion to Britain and additionally some of the most famous businesses were founded by Jews.

---

24 cf. http://bit.ly/1GdU1N8 (02-11-2014)
25 cf. Winder, pp.226-229
26 cf. ibid. pp.231-236

1.3.3 The Huguenots

Prosecution of French Huguenots started early in the 17th century, when Cardinal Richelieu decided to erase the protestants from his catholic country. In 1685, the protestant services were prohibited, the churches demolished and trade was restricted. Children should be baptised and raised as Catholics. There was only one option for those who didn't want to suffer through this – escaping to England, which meant more than just taking the next ship. There were patrols on the beach, some families were observed. It was indeed very difficult to actually get on a ship, and when they managed this they had to hide under straw or coal or in empty wine barrels. The reasons why Charles the second gave the French permission to enter England were easy. France was an enemy, and every enemy's enemy is a friend of of them. The Great Plague in the 1660s was a motive as well because the Country was lacking in workers.[27] Word of the fugitives spread, and the British population wanted to help. They raised money, the French churches in London supported them and to counteract starvation, they set up soup kitchens.[28]

And the French people succeeded in Britain:

> *"The Huguenots possessed exactly what the country needed: the know-how necessary to transform an agricultural economy into an industrial one. They became spinners in Bideford, tapestry weavers in Exeter and Mortlake,[...]. Featherwork, fans, girdles, needles, soap, vinegar – whole new trades sprang into existence. They revolutionised the silk industry, and brought new techniques for velvet, taffeta and brocade.[29]*

They brought their very own vivid traditions and handicraft to England and the nation became one of the major textile export countries.[30] London's Spitalfields became 'weaver town' because the weaving industry flourished under the Huguenots. When the silk industry receded, the French settled into London's suburbs.[31]

---

27 cf. Winder, pp.78-81
28 cf. ibid. pp.81-82
29 ibid. p.82
30 cf. ibid. p.83
31 cf. http://bbc.in/10eoRnH (02-11-2014)

2. The Cultural Heritage of the Immigrants

2.1 The International Influence on Food

Going out to get Dinner in London is an endless struggle. There are too many options, too many restaurants and takeaways to choose from. Do you want to get street food or go to a Michelin-Star restaurant? Do you want Indian, Chinese, Italian or Japanese food? In London, you can have everything.

Back in 1957, only fifty Chinese Restaurants existed in the whole country of Great Britain. 1963, only six years later, there were 1500.[32] Nowadays they're nearly uncountable, because there's always a new restaurant opening and another one closing its doors forever.

The main reason for the variety of food you can get in London, and it doesn't matter if it's from a small booth in Camden Market or from a five Star exclusive Restaurant in Kensington, is that all Immigrants had the need to surround themselves with the food from their homelands. Consequently they set up restaurants and shops to provide them with their local dishes. These influences became more and more important for the British population, as Britain grew to be more and more dependent on imports. Also, the population became wealthier which meant that they had access to a bigger assortment of products, which included not only national products, but also international and ethnic food.[33]

This leads to the topic of cookbooks, because the population now had access to international products, but they didn't really know how to use and cook them.

After Panikos Panayi, you can divide English cookbooks into three categories:

> "[...]those which have overtly made a case about their Englishness or Britishness; 'general cookbooks' without any claims about the national origins of the food concerned; and those which look at the cuisines of specific countries or ethnicities. The last group, [....], has focused especially on Indian, Chinese and Italian food." [34]

This kind of division makes it difficult for readers to choose what they want.

---

32 cf. Winder, *Bloody Foreigners*, picture 17 between p.282 and p. 283
33 cf. Panayi, *Spicing Up Britain*, p. 9
34 Panayi, p.14

A British housewife for example would want to have a cookbook which would show her the good traditional fare. Tourists would long for a book that would give them as much recipes from their trip as possible, which would definitely include some foreign dishes, as they probably didn't eat in pubs all the time. The different needs of different people would ultimately explain the different kinds of cookbooks in England.

There are many stereotypes about the original British kitchen, some of them say it's tasteless and dull. But you can't open a "British" cookbook these days without finding at least one Curry recipe in it - but Curry comes from India, doesn't it?

Yes, it does and it's definitely not tasteless, as the Indian kitchen uses much spices. But it's still not British. So why is it in a British cookbook?

Rose Prince, author of the highly praised "New English Kitchen" states it like that:

> "The national cuisine may be fossilised in people's minds as pies, roasts and nursery puddings, but there is now no reason why it could not include the rice noodle dishes of Southeast Asia or the delicious food of the Mediterranean. This is after all a country with a five-hundred-year-old history of food piracy: borrowing ideas from other shores, importing their raw materials and learning to cultivate them in our soil."[35]

She's principally saying that it's totally okay to include food with non-British origin in cookbooks and to appreciate the cultural influence from other countries.

That's why, if you ask people what their favourite food on their London Trip was, often the answer is not the typical "Fish-and-Chips" it's more often "a good curry on Brick Lane" or "the best Chow Mein of my life in Chinatown". This is exactly what it's all about. Immigrants taking their national foods with them and introducing them to the world. That's the answer to the question why there's curry in English cookbooks. A walk in Brick Lane can take you straight to Mumbai.

And, to end this chapter, the presumably most British dish ever, Fish-and-Chips, has got foreign origins itself. Fried fish probably developed from Jewish traditions and chips originate from France.[36]

---
35 Panayi, p.13
36 cf. Panayi, p.17

2.2 Religion- The Brick Lane Mosque and The Neasden Temple

Religion has played and still plays a big role for a lot of people around the world. Immigrants often adapted to the religion in their new country, but a lot also kept their own and practised them in their new home.

One can see this in the manifestations they built for their faith - temples and mosques, churches and chapels. Presently, you can find 373 Mosques in London. [37] Berlin for example, who has got a rather large Muslim community itself got only around 80 mosques.[38]

One rather large Mosque In London which is also connected with a lot of Immigration History is the "Brick Lane Jamme Masjid", a Mosque located on the corner of Brick Lane and Fournier Street in London's East End.[39]

(http://bit.ly/1wDiOF4, 31-10-2014)

The ordinary looking House, built in 1743, became the "New French Church" of the Huguenots, the French protestants who fled their mother country because they were religiously persecuted. In 1819, it became a Methodist Chapel, but yet again it didn't stay like this for long because in 1897 the Chapel was converted by Jews and was now known as the "Spitalfields Great Synagogue". It remained like this until the 1960s, when the Jews located in London's East End all moved to London's northern suburbs.[40]

---

37 cf. http://bit.ly/10yHa86, 29-10-2014
38 cf. http://bit.ly/1tikOfy, 29-10-2014
39 cf. http://bit.ly/1wDiOF4, 29-10-2014
40 cf. http://bit.ly/1ubUoEF, 31-10-2014

1976, the Synagogue opened its doors again-but this time as a mosque for the newly arrived Muslim migrants from eastern India and Bangladesh.[41]

Unlike the Brick Lane Mosque, the Neasden Temple located in the North-West of London, is open to public. [42]

(http://bit.ly/1q7xmrr, 31-10-2014)

The "BAPS Shri Swaminarayan Mandir" is one of the biggest Hindu Temples outside India. The construction of the magnificent Mandir began in 1992 and it took only three years to build it, for it was inaugurated in 1995.[43] Since then it attracted more than 5 million visitors. [44] The whole building is made without any steel amplifications, the only construction materials used by over 1000 volunteer worker were 3000 tons of Bulgarian limestone, 1200 tons of Italian marble and another 900 tons of Indian marble. The stones were hand-carved in India, then sent to Britain and finally assembled to create one large temple.[45] Jessica Cargill Thompson from the website "Time Out" says: "[...], it is impossible to stand here without feeling spiritually moved and inwardly contemplative."[46]

---

41 cf. http://bit.ly/1ubUoEF, 31-10-2014
42 cf. http://bit.ly/1sPEwjZ, 31-10-2014
43 cf. http://bit.ly/1u02kbz, 31-10-2014
44 cf. http://bit.ly/1nmKc56, 31-10-2014
45 cf. http://bit.ly/1zR4z3Z, 31-10-2014
46 http://bit.ly/1u02kbz, 31-10-2014

Conclusion

Immigrants have changed London and London has changed the Immigrants. They brought their own culture, their food, their traditions and they are not isolated as one could think. They assimilate with the city and the city adjusts to their needs.
London would not be the same without them. Racial Discrimination and Prejudice are not as big a problem as in many other immigrant cities.
The new big group of Immigrants coming to Britain are Italians, Greeks and Spanish from South-Europe suffering from the recession.[47] Let's hope the city will welcome and incorporate them like every other group of Immigrants!

> *"With more inter-marriage, more mixed communities, more ethnically mixed children and more diversity, London is set in the 21st century to become a new type of city for Europe - one more like the immigrant cities of the United States, but without, if London gets it right, their segregation."*[48]

---

47 cf. http://bit.ly/1hrvyqN (02-11-2014)
48 http://ind.pn/1A0NWDg (02-11-2014)

Bibliography:

Books:

Bromhead, Peter, *Life in Modern Britain,* 1982[7]

Panayi, Panikos, *Spicing up Britain-The Multicultural History of British Food,* London, 2008[1]

Winder, Robert, *Bloody Foreigners- The Story of Immigration to Britain,* London, 2005[1]

Websites:

http://www.bbc.co.uk/history/british/empire_seapower/east_india_01.shtml (02-11-2014)

http://www.bbc.co.uk/legacies/immig_emig/england/london/article_1.shtml (02-11-2014)

http://www.bbc.co.uk/london/content/articles/2005/05/26/indian_london_feature.shtml (02-11-2014)

http://www.bbc.co.uk/london/content/articles/2005/05/27/chinese_london_feature.shtml (02-11-2014)

http://www.bricklanejammemasjid.co.uk/ (02-11-2014)

http://www.english-heritage.org.uk/caring/listing/heritage-centenary/landmark-listings/brick-lane-jamme-masjid (02-11-2014)

http://www.globalirish.ie/issues/how-many-irish-people-live-abroad-an-ean-factsheet/ (02-11-2014)

http://www.independent.co.uk/news/uk/london-europes-new-ethnic-melting-pot-1525506.html (02-11-2014)

http://londonmandir.baps.org/ (02-11-2014)

http://londonmandir.baps.org/images/the-mandir-moods/ (02-11-2014)

http://londonmandir.baps.org/the-mandir/mandir-how-it-was-made/ (02-11-2014)

http://migrationobservatory.ox.ac.uk/briefings/migrants-uk-overview (02-11-2014)

http://www.migrationobservatory.ox.ac.uk/briefings/london-census-profile (02-11-2014)

http://mosques-map.muslimsinbritain.org/maps.php#/town/London (02-11-2014)

http://www.nationalarchives.gov.uk/pathways/citizenship/brave_new_world/immigration.htm (02-11-2014)

http://news.bbc.co.uk/hi/english/static/in_depth/uk/2002/race/short_history_of_immigration.stm#1500 (02-11-2014)

http://news.bbc.co.uk/2/shared/spl/hi/uk/05/born_abroad/countries/html/republic_of_ireland.stm (02-11-2014)

http://www.oldbaileyonline.org/static/Irish.jsp (02-11-2014) http://www.taz.de/!77770/ (02-11-2014)

http://www.telegraph.co.uk/expat/expatpicturegalleries/9673025/Londons-immigrant-districts.html?frame=2396309 (02-11-2014)

http://www.telegraph.co.uk/news/picturegalleries/uknews/8296462/Chinese-New-Year-Londons-Chinatown-prepares-for-Year-of-the-Rabbit.html (02-11-2014)

http://www.telegraph.co.uk/news/uknews/immigration/10480785/Most-immigrants-to-the-UK-now-come-from-China-figures-show.html (02-11-2014)

http://www.theguardian.com/commentisfree/2009/jan/06/religion-christianity (02-11-2014)

http://www.theguardian.com/commentisfree/2013/jan/28/british-dream-europe-african-citizens (02-11-2014)

http://www.theguardian.com/uk-news/2014/feb/27/net-migration-uk-jumps-30-percent (02-11-2014)

http://www.timeout.com/london/things-to-do/seven-wonders-of-london-baps-shri-swaminarayan-hindu-mandir (02-11-2014)

http://www.visitjewishlondon.com/uk-jewish-life/history (02-11-2014)

YOUR KNOWLEDGE HAS VALUE

- We will publish your bachelor's and master's thesis, essays and papers

- Your own eBook and book - sold worldwide in all relevant shops

- Earn money with each sale

Upload your text at www.GRIN.com
and publish for free